BUILDING NUMBER SENSE with TEN-FRAMES

This book belongs to:

number +

www.numberplus.co.uk

First published 2025
001

Copyright © Paula Claughton, 2025
All images copyright © Number Plus, 2025

The moral right of the author has been asserted.

All rights reserved. No part of this publication may be reproduced, stored in a retrieval system or transmitted in any form or by any means, electronic, mechanical, photocopying, recording, or otherwise, without prior permission in writing from the publisher.

Illustrations by Marco Zerneri

paula.claughton@numberplus.co.uk

Scan the QR code
to download bonus worksheets

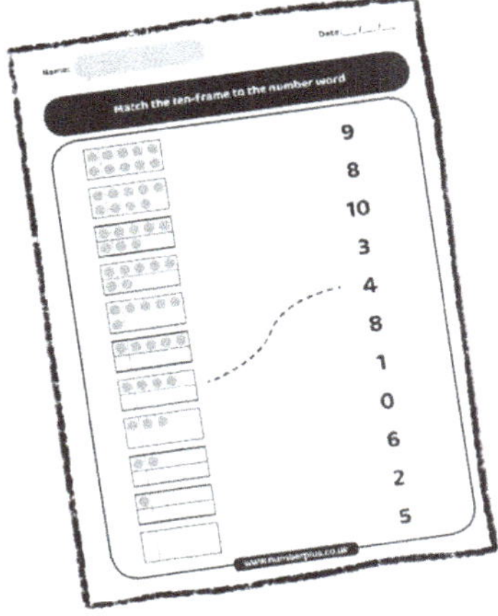

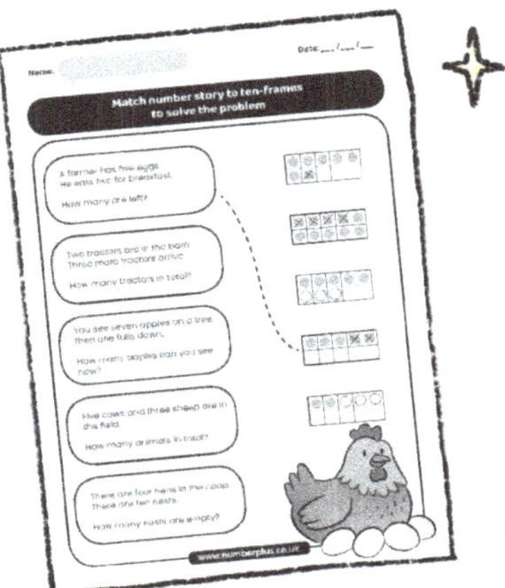

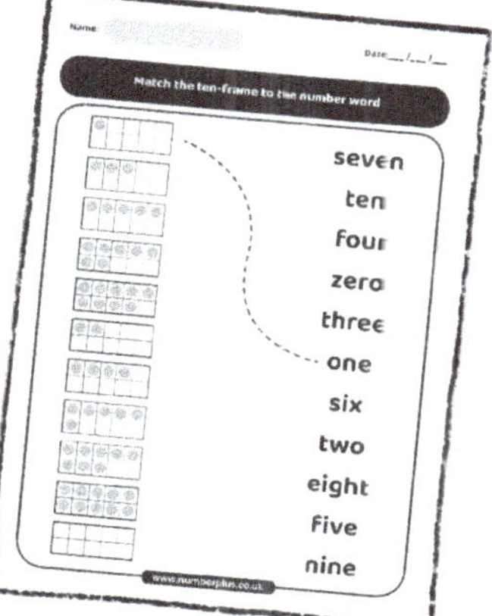

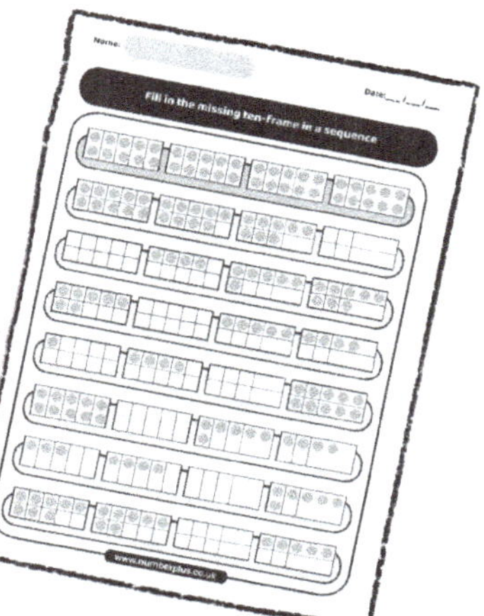

Contents

What is a ten-frame?

A ten-frame is a simple maths tool that helps young children **see** and **understand** numbers.

It looks like a rectangle with 10 boxes - two rows of five.

It helps children quickly see how many items are in a group without having to count them one by one.

Example: 3

It shows missing pieces. If some boxes are empty, children can figure out how many more you need to make 10.

Example: 5 + ☐ *= 10*

Why is a ten-frame helpful?

Ten-frames are great for little learners because they:

- Help children spot patterns, like 5 and 5 make 10
- Make it easier to add and take away
- Build a strong sense of how numbers work
- Help children recognise numbers by sight

How does this book help my child?

We use a simple three-step method to help your child understand numbers deeply. It's called **Concrete–Pictorial–Abstract (CPA)**:

- **Concrete** Children play with real objects like blocks or buttons to show numbers.
- **Pictorial** They use pictures or drawings to see the numbers in a new way.
- **Abstract** They start to use number symbols (like 5 or +) to solve problems.

This step-by-step approach helps children move from playing with objects to thinking with numbers.

To make things easier for young children, we've used simple, child-friendly words instead of tricky maths terms.

This key shows you the words we've chosen to help your child understand as they learn:

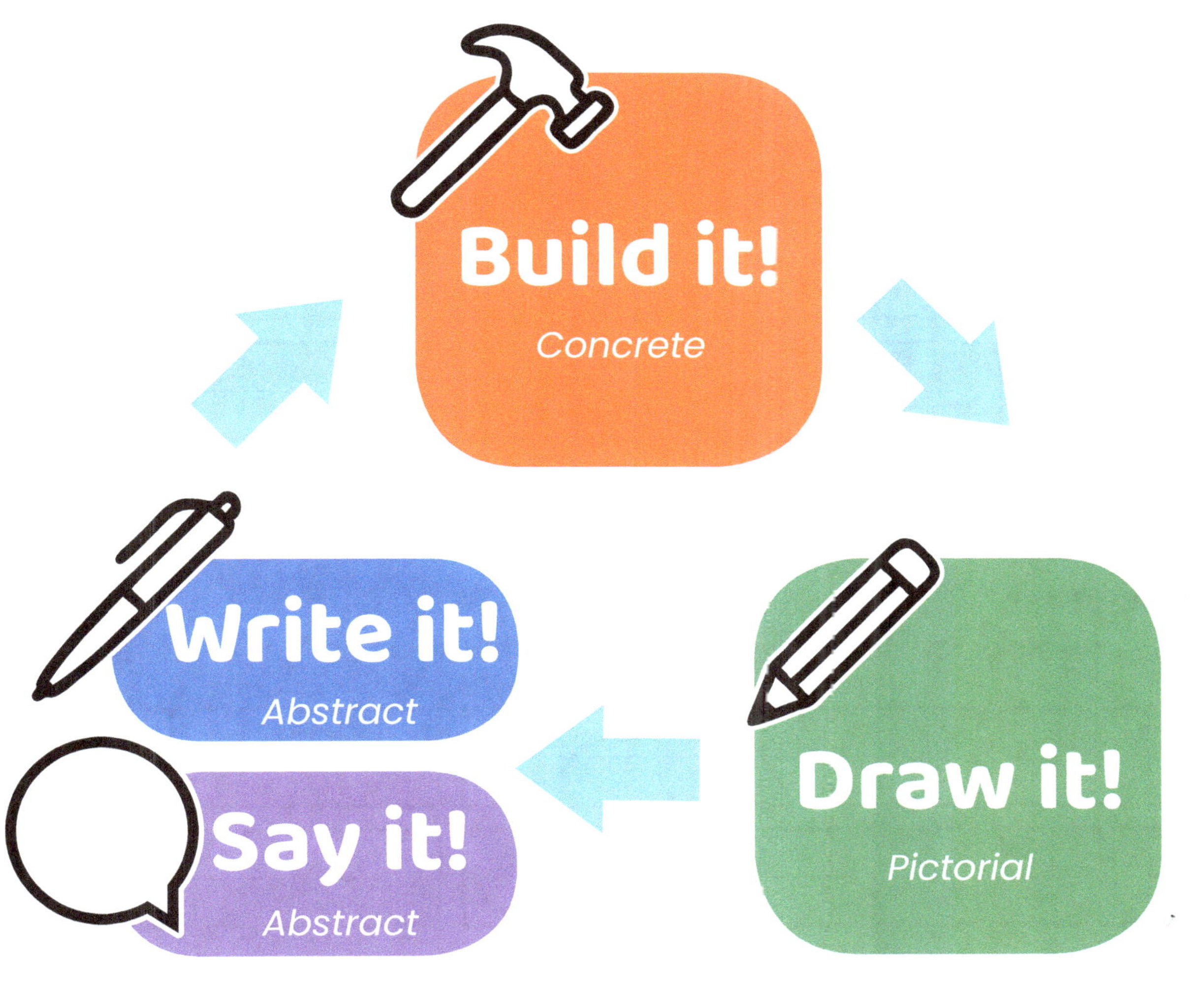

Spotting Number Patterns

Ten-frames help children build early number skills by showing numbers in simple patterns. Using different patterns on a ten-frame helps children build number sense and notice important relationships. *For example:*

2s Pattern
Place counters in pairs—one on top, one below.

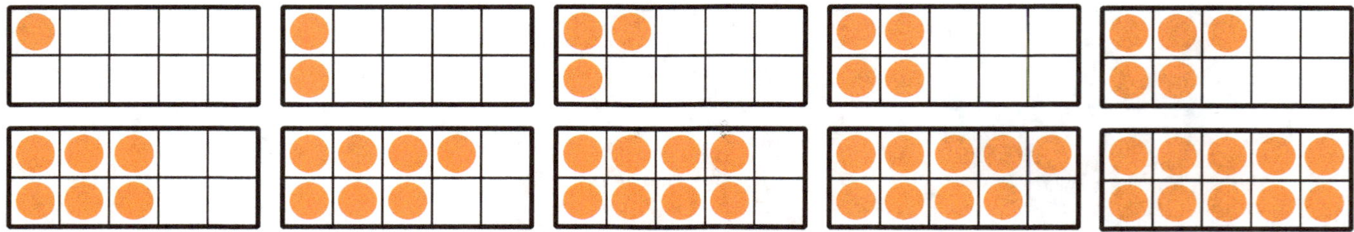

This helps children spot even and odd numbers and practise counting in twos.

5s Pattern
Place objects across the top row from left to right, then continue on the bottom row.

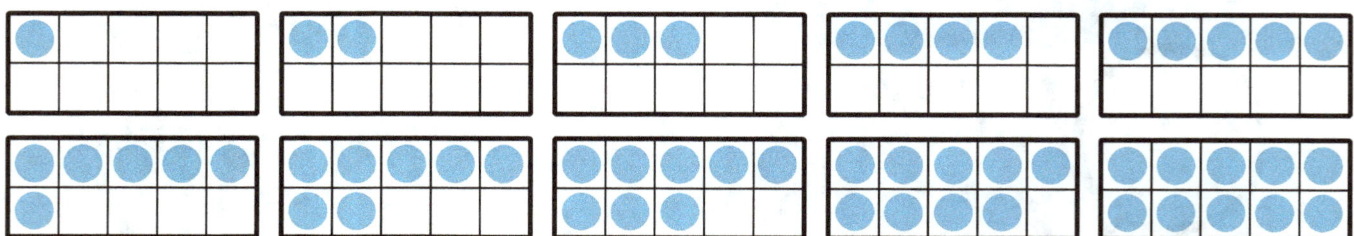

This helps children see how numbers are grouped in 5s and 10s.

There's no one "right" way to use a ten-frame, what matters is helping children make sense of numbers. Encourage them to try both patterns and notice what works best for them.

Finger Patterns
Young children love using their fingers, and they're a brilliant maths tool!

0

Finger patterns help children:
- See numbers quickly without counting
- Understand how numbers are made (*e.g. 3 and 2 make 5*)
- Build confidence and fine motor skills

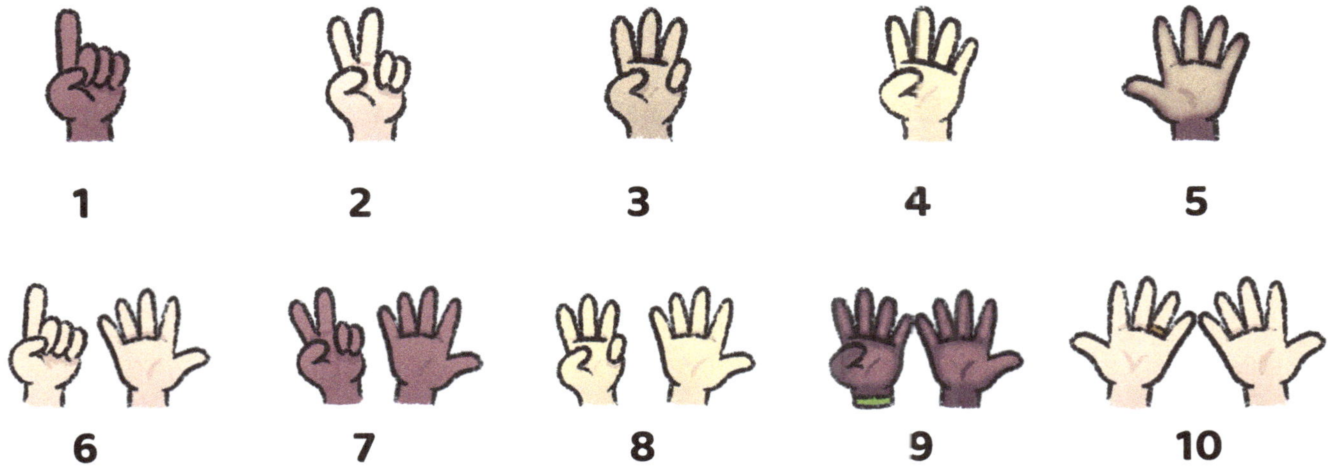

1   2   3   4   5

6   7   8   9   10

Look for this icon in activities.

It's a friendly reminder to use fingers alongside the ten-frame!

What do you notice?

This book gently changes how problems are shown, so your child can spot patterns and think flexibly.
For example, children might:

Build 5 on a ten-frame using objects

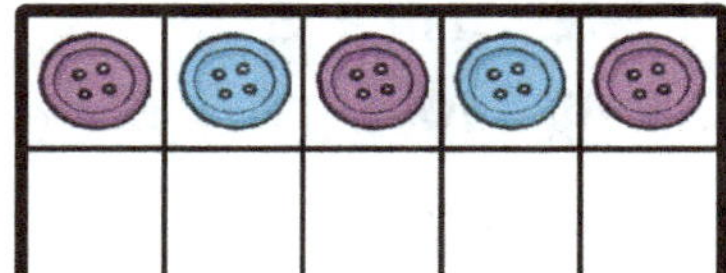

Draw 5 dots on a ten-frame

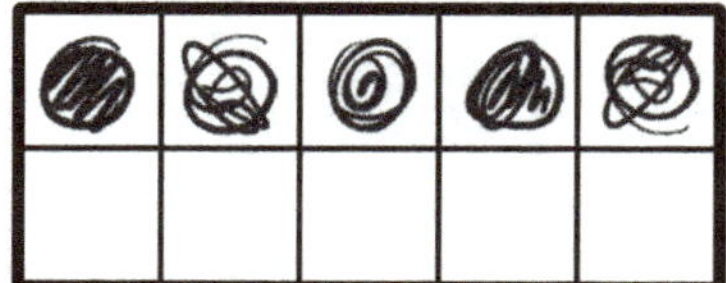

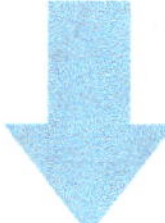

Show 5 as a finger pattern

Also, by carefully changing the steps, children start to notice what stays the same and what's different, helping them make sense of how maths works.

Example:

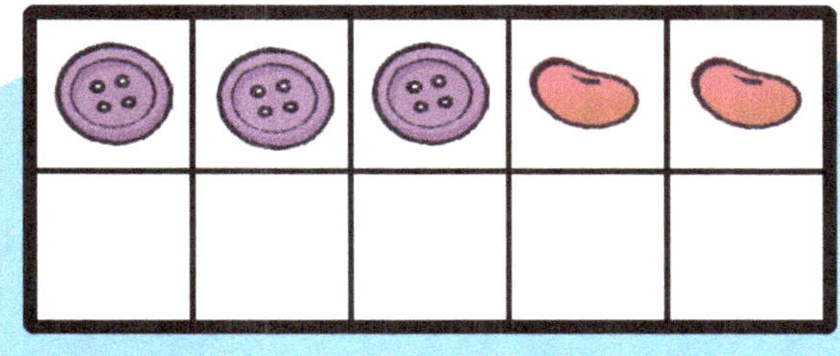

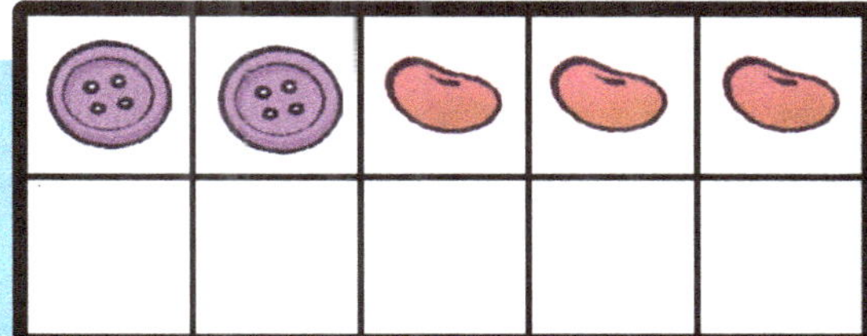

$$3 + 2 = 5 \qquad 2 + 3 = 5$$

?

What's the same?
Both make 5.

?

What's different?
Numbers are
in a different order.

This helps build number sense, an understanding of how numbers work and relate to each other.

To find out more, please go to the 'Answers Explored' section at the back of this book

Let's talk about it!

Talking about maths helps children understand numbers better.

When we explain our thinking, our brains get stronger, and we learn new ways to solve problems.

Ten-frames help children talk about maths by giving them a clear way to see and explain numbers, patterns and relationships, such as:

You can help by asking questions like:

What's the same?

What do you see?

How did you work that out?

What's different?

What happens if we add one more?

These **little** chats help children grow **big** maths ideas!

Helpful Hints

- Sit together and enjoy the number stories.

- Use small objects (buttons, cereal) as counters.

- Ask simple questions like 'Can you show me 5?' or 'How many do you have?'

- Let them notice patterns like rows of 5 on a full ten-frame.

- It's okay if they make mistakes, mistakes help us learn!

- Cheer them on. Encouragement goes a long way.

Getting started

Turning everyday objects like empty egg boxes, buttons, or snacks into maths tools helps children make connections between numbers and the real world. It also brings maths to life in a fun and meaningful way so, let's get started!

First, cut out the ten-frame templates from the back of this book OR...build your own using everyday objects.
Here are a few ideas to build ten-frames with children...

Then, cut out the counters at the back of this book OR...gather some small objects which are the same or similar size, shape and colour, like...

Buttons

Coins

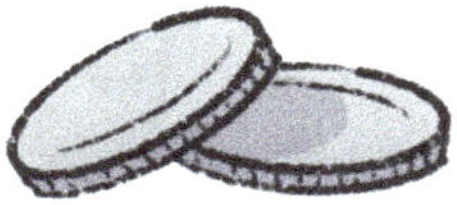

Small Toys

Dried pasta

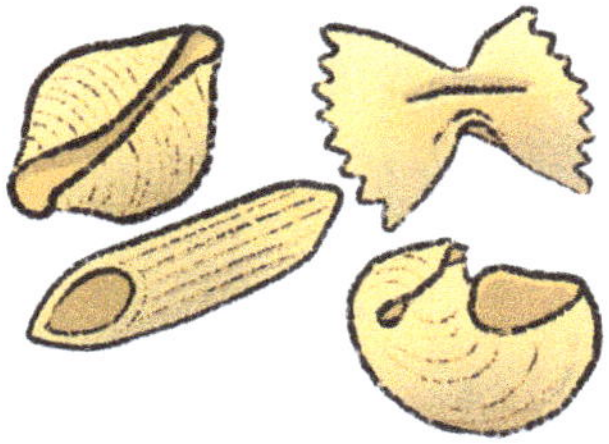

Pebbles

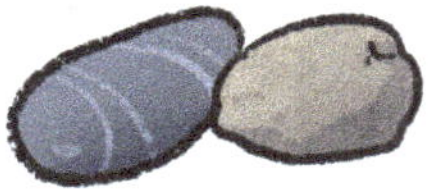

Shells

Now, have fun playing the **'Build It!'** activities with young learners.

Top Tip: *Laminate or glue resources onto card so they last longer.*

Build it!

Have ready:

- 1 empty ten-frame
- objects or counters (10)
- number cards (0-10)

Optional:

- building blocks (such as, Lego, Mega Bloks, wooden cubes)
- healthy snacks (for example, grapes, crackers, cereal)
- chalk or tape

This step supports the **concrete** stage of learning, which is essential at this age. Using real objects helps children build strong mental images of numbers they can picture and remember later.

Your child might:
Fill one row before starting the next (to show 5s and 10s),

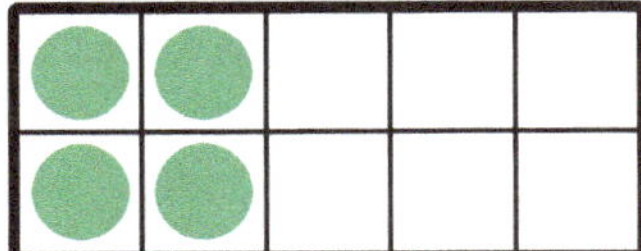

Make vertical pairs (to show 2s and even numbers),

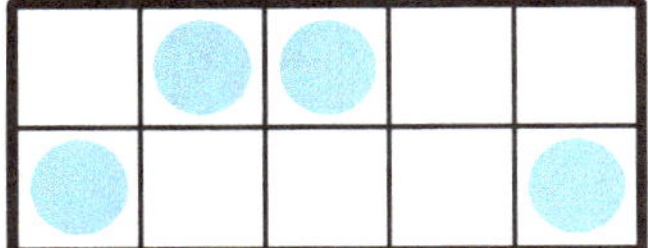

Or use shapes and colours to build their own patterns.

Important Point
While exploring is great, try encouraging structured ones (like filling from the top left) so number patterns are clearer and easier to talk about.

We've included 12 fun activities to help your child get used to using ten-frames. These games are a super way to help them see, touch, and build numbers, right from the start!

1. Exploring the Ten-Frame

Give the child an empty ten-frame and some counters or objects (for example, buttons, beads or blocks). Let them freely explore placing objects in the spaces, counting as they go.

This activity encourages curiosity and builds confidence, making children more open to exploring other maths concepts.

2. Fill the Frame

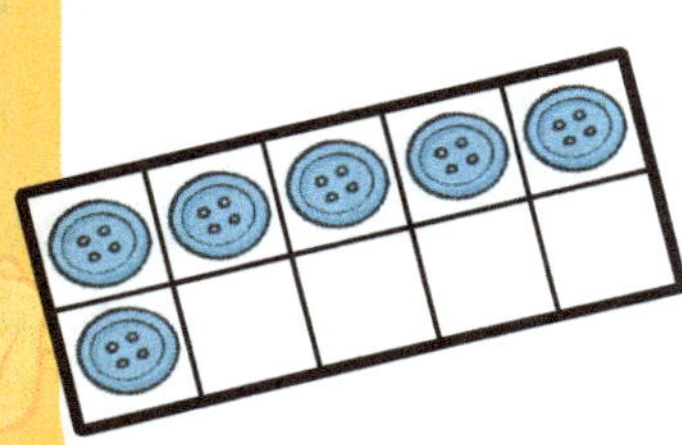

Give the child a number (for example, 6). Have them place 6 small objects on a ten-frame.
Ask: 'How many more do we need to make 10?' Let the child add the missing objects. Repeat with a different number.

Filling a ten-frame offers a hands-on, visual way for children to explore and understand numbers.

3. One More, One Fewer

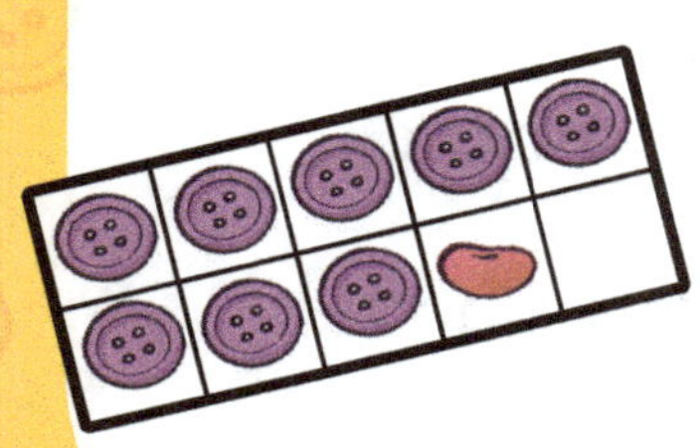

Start with a certain number of objects on a ten-frame (for example, 8). Ask: 'What happens if we add one more?' or 'What if we take one away?' Have child adjust their frame and count.

This helps children compare and build number sense.

4. Counting Up and Down

Start with a small number of objects (for example, 4) on a ten-frame. Add objects one by one while counting up (for example, 4, 5, 6), then take them away while counting down (for example, 6, 5, 4).

This helps children practise counting forwards and backwards in ones.

5. Ten-Frame Flash

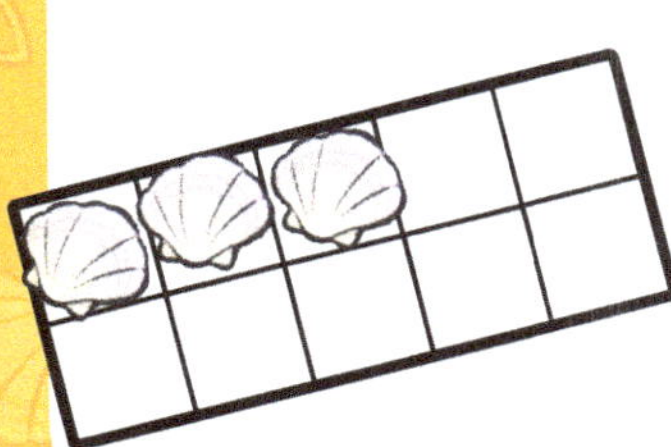

Whilst the child closes their eyes, build a number with objects on a ten-frame (for example, 3). Show the ten-frame for 3 seconds and cover it. Ask: 'How many without counting?' Uncover the ten-frame to check. Repeat with a different number of objects.

This helps with quick number recognition!

6. Snack Attack

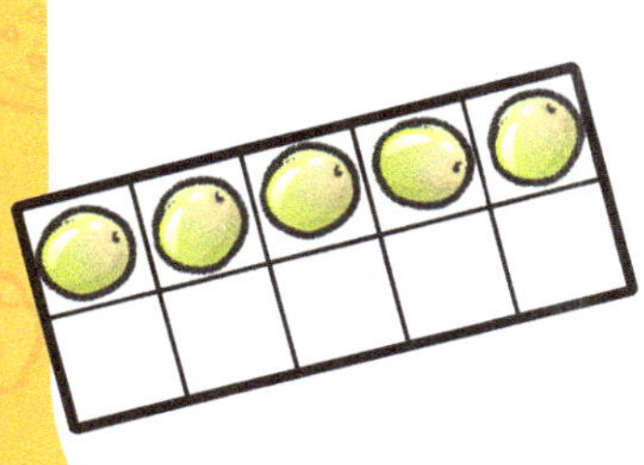

Use healthy snacks like grapes, cereal pieces or crackers as counters. Hold up a number card (0–10). Child places that many snacks on a ten-frame (for example, 5) and counts backwards each time a snack is eaten.

Using tangible snacks to fill the ten-frame allows children to physically manipulate objects while counting.

7. Ten-Frame Towers

The child places objects on a ten-frame, then builds a tower with the same number of blocks.

This activity enables children to visually compare quantities.

8. Match My Number

Hold up a number card (0-10). The child places that many objects on a ten-frame to match the number.

This activity helps children connect a number to a quantity.

9. Shake and Roll

Have ready 10 objects. Without looking, the child grabs a small number of objects, shakes their hand and rolls the objects onto a flat surface.
Ask: 'How many do you have?' Check, by placing the objects onto a ten-frame.

This activity encourages children to estimate.

10. Move and Count

Create a giant ten-frame on the floor (using chalk or tape). The child becomes the counters, stepping into the squares as they count aloud.

Jumping into drawn squares connects physical movement with counting, reinforcing number order in a dynamic way.

11. Ten-Frame Treasure Hunt

Hide objects around the room. The child finds and places them one by one onto a ten-frame, counting as they go.

This activity reinforces the concept of one-to-one correspondence.

12. Clap and Fill

Clap a number between 1 and 10. The child places that many objects on a ten-frame and finds the number card that matches the quantity.

This activity reinforces the link between sounds and numerical quantities.

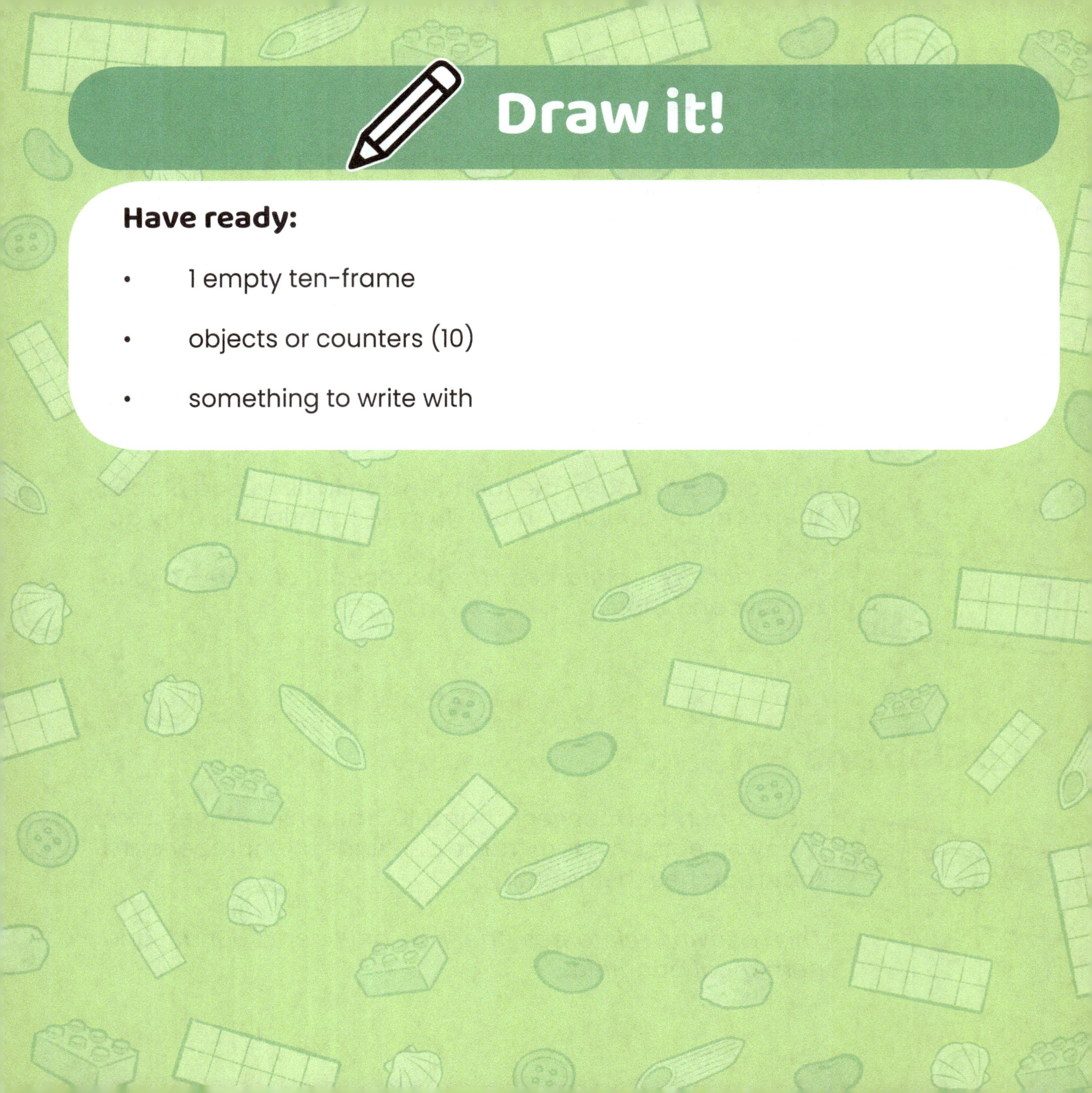

Draw it!

Have ready:

- 1 empty ten-frame
- objects or counters (10)
- something to write with

Draw it!

This step is all about visualising numbers—it's part of the **pictorial** stage of learning. First, children use real objects (the concrete stage), and then they show what they know by making pictures. This helps bridge the gap between hands-on experiences and thinking in symbols.

Example: To show 4, your child might draw 4 dots in a ten-frame picture.

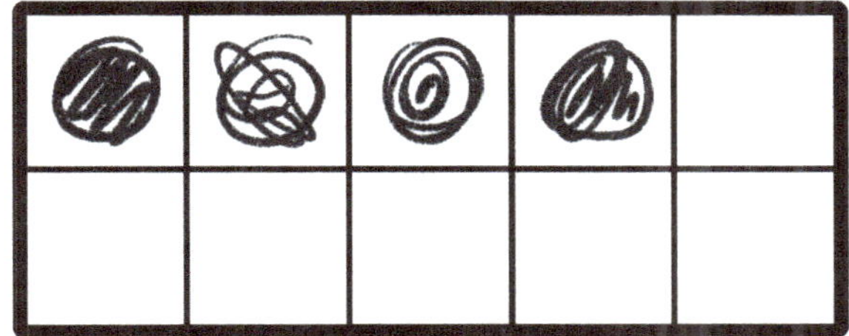

Drawing like this helps children picture numbers in their heads, even when the real objects aren't there.

Each activity in this section includes a simple number story, because linking maths to real-life situations helps young children understand what numbers really mean.

But just as important as doing the activity is **talking** about it together. Chatting about numbers—what your child sees, thinks, and notices—helps them build confidence and make sense of how numbers work. Whether you're counting eggs in a basket or working out how many cupcakes have been eaten, these **little** conversations help your child grow **big** maths ideas!

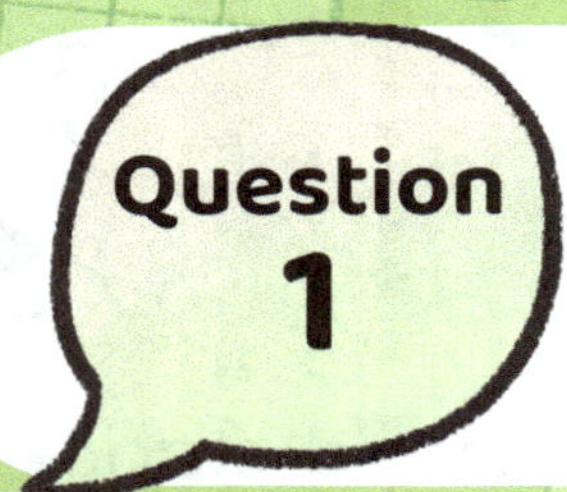

A hen laid 3 eggs in the morning.
In the afternoon, she laid 4 more.
How many eggs are there now?

🔨 Build it!

First, place 3 objects on a ten-frame. Then add 4 more.

✏️ Draw it!

Then, draw 3 dots on the ten-frame. Add 4 more.

🔍 Solve it!

There are ⬚ eggs now.

Charlie baked 4 cupcakes and put them on a tray.
Then he made 3 more.
How many cupcakes does Charlie have now?

 **Build it!**

First, place 4 objects on a ten-frame. Add 3 more.

Draw it!

Then, draw 4 dots on the ten-frame. Add 3 more.

Solve it!

Charlie has ______ cupcakes now.

A chicken farmer packed 6 eggs in a basket, but 3 eggs cracked.
How many eggs did not crack?

 **Build it!**

First, place 6 objects on a ten-frame. Remove 3.

Draw it!

Then, draw 6 dots on the ten-frame. Cross out 3.

 **Solve it!**

☐ eggs did not crack.

Kai had 6 cupcakes on a tray.
He gave 2 cupcakes to his friends.
How many cupcakes were left on the tray?

 **Build it!**

First, place 6 objects on a ten-frame. Remove 2.

Draw it!

Then, draw 6 dots on the ten-frame. Cross out 2.

 **Solve it!**

cupcakes were left.

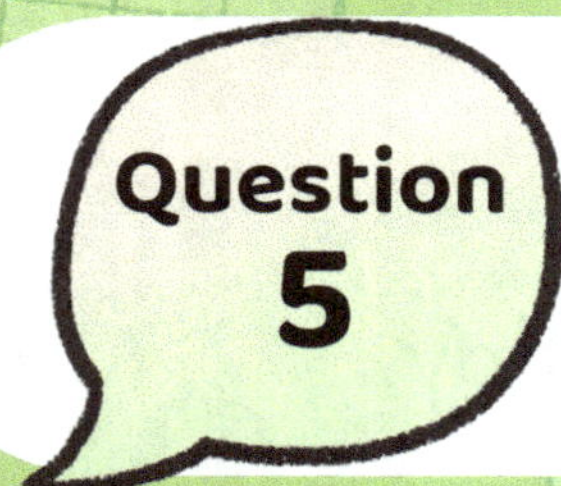

The farmer has 6 cows in the barn.
He wants to make it 10 cows.
How many more cows does the farmer need?

 **Build it!**

First, place 6 objects on a ten-frame.
Add more until the total makes 10.

 **Draw it!**

Then, draw 6 dots on the ten-frame. Add more until there are 10.

 **Solve it!**

The farmer needs [ ] more cows.

A baker made 10 cupcakes in the morning.
4 were sold.
How many cupcakes are left?

 **Build it!**

First, place 10 objects on a ten-frame. Remove 4.

Draw it!

Then, draw 10 dots on the ten-frame. Cross out 4.

Solve it!

There are ☐ cupcakes left.

There are 9 chickens in the barn.
3 chickens run out of the barn.
How many chickens are in the barn now?

🔨 Build it!

First, place 9 objects on a ten-frame, remove 3, and count how many are left.

✏️ Draw it!

Then, draw 9 dots on the ten-frame. Cross out 3.

🔍 Solve it!

There are ☐ chickens in the barn.

Freya baked 8 cupcakes, but 2 got eaten. How many cupcakes have not been eaten?

Build it!

First, place 8 objects on a ten-frame. Remove 2.

Draw it!

Then, draw 8 dots on the ten-frame. Cross out 2.

Solve it!

___ cupcakes have not been eaten.

In an egg box, there are 2 big eggs and 3 small eggs.
How many eggs are in the box altogether?

🔨 Build it!

First, place 2 objects on a ten-frame, then add 3 more.

✏️ Draw it!

Then, draw 2 big dots and 3 small dots on the ten-frame.

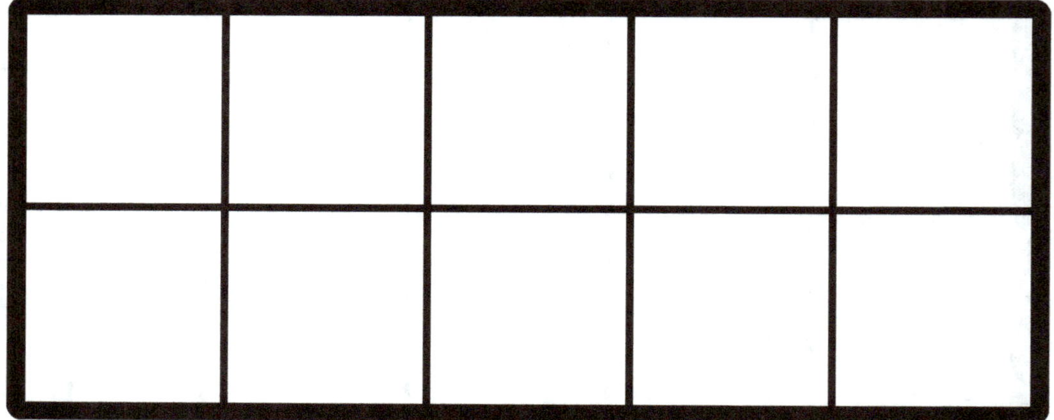

🔍 Solve it!

There are ☐ eggs in the box.

Amina baked 1 cupcake first and then made 4 more. How many cupcakes did Amina bake altogether?

🔨 Build it!

First, Place 1 object on a ten-frame. Add 4 more.

✏️ Draw it!

Then, draw 1 dot on the ten-frame. Add 4 more.

🔍 Solve it!

Amina baked ⬜ cupcakes altogether.

Write it!

Have ready:

• 1 empty ten-frame

• objects or counters (10)

• something to write with

Write it!

By this point, children have already explored numbers using real objects (*concrete*) and pictures (*pictorial*).

Now, in the '*Write it!*' step, they build on what they know by writing numbers and words so that number sentences (*like 2 + 3 = 5*) start to make sense.

This is part of the **abstract** stage of learning and helps them connect what they do and see with what they write and understand.

Each page includes a fun bus-themed story to bring numbers to life. These stories help your child see why numbers matter in the real world. Plus, the repeated structure builds confidence and encourages independent thinking.

And just like every part of this resource, these activities are full of chances to talk together about numbers.

Chatting about what's happening, what your child notices, or what might happen next helps them develop deeper thinking and builds their maths confidence!

🔨 **Build it!**   First, build 1 on a ten-frame with real objects.

✏️ **Draw it!**   Then, draw 1 dot on the ten-frame.

1 one

🖊️ **Write it!**   Now, write 1 as a number and a word.

First School Bus

First, there were 2 people on the bus.
Then, 1 gets off.

How many people are on the bus now?

🔍 **Solve it!**

Now, there is ⬜ person on the bus!

Build it!
First, build 2 on a ten-frame with real objects.

Draw it!
Then, draw 2 dots on the ten-frame.

2   two

Write it!
Now, write 2 as a number and a word.

Second School Bus

First, 1 person gets on the bus.
Then, 1 more gets on.

How many people are on
the bus now?

Solve it!

Now, there are ___ people on the bus!

Build it! First, build 3 on a ten-frame with real objects.

Draw it! Then, draw 3 dots on the ten-frame.

3 three

Write it! Now, write 3 as a number and a word.

Third School Bus

First, there were 6 people on the bus.
Then, 3 get off at the next stop.

How many people are on the bus now?

Solve it!

Now, there are ☐ people on the bus!

Build it! First, build 4 on a ten-frame with real objects.

Draw it! Then, draw 4 dots on the ten-frame.

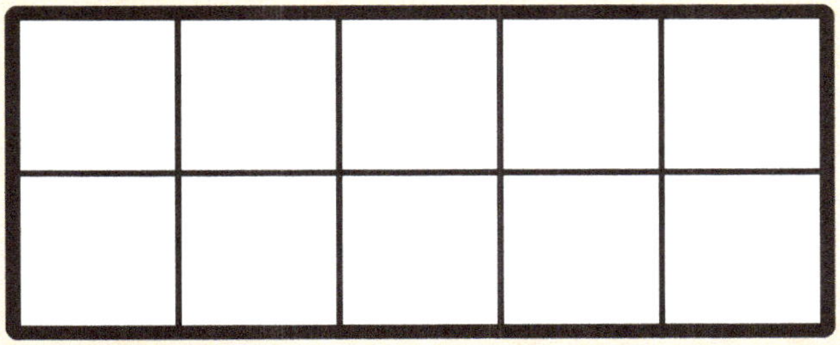

4   four

Write it! Now, write 2 as a number and a word.

Fourth School Bus

First, there were 2 people on the bus.
Then, 2 more get on at the next stop.

How many people are on the bus now?

Solve it!

Now, there are ☐ people on the bus!

🔨 **Build it!**   First, build 5 on a ten-frame with real objects.

✏️ **Draw it!**   Then, draw 5 dots on the ten-frame.

5 five

✒️ **Write it!**   Now, write 5 as a number and a word.

Fifth School Bus

First, there were 10 people on the bus.
Then, 5 get off.

How many people are on the bus now?

🔍 **Solve it!**

Now, there are ☐ people on the bus!

Build it! First, build 6 on a ten-frame with real objects.

Draw it! Then, draw 6 dots on the ten-frame.

6   six

Write it! Now, write 6 as a number and a word.

Sixth School Bus

First, there were 3 people on the bus.
Then, 3 more get on.

How many people are on the bus now?

Solve it!

Now, there are [ ] people on the bus!

Build it!

First, build 7 on a ten-frame with real objects.

Draw it!

Then, draw 7 dots on the ten-frame.

7 seven

Write it!

Now, write 7 as a number and a word.

Seventh School Bus

First, there were 8 people on the bus.
Then, 1 gets off.

How many people are on the bus now?

Solve it!

Now, there are ☐ people on the bus!

Build it! First, build 8 on a ten-frame with real objects.

Draw it! Then, draw 8 dots on the ten-frame.

<table>
<tr><td></td><td></td><td></td><td></td><td></td></tr>
<tr><td></td><td></td><td></td><td></td><td></td></tr>
</table>

8 eight

Write it! Now, write 8 as a number and a word.

Eighth School Bus

First, there were 4 people on the bus.
Then, 4 get on.

How many people are on the bus now?

Solve it! Now, there are ☐ people on the bus!

Build it!

First, build 9 on a ten-frame with real objects.

Draw it!

Then, draw 9 dot on the ten-frame.

9 nine

Write it!

Now, write 9 as a number and a word.

Ninth School Bus

First, there were 10 people on the bus.
Then, 1 gets off.

How many people are on the bus now?

Solve it!

Now, there are [ ] people on the bus!

🔨 **Build it!** First, build 10 on a ten-frame with real objects.

✏️ **Draw it!** Then, draw 10 dot on the ten-frame.

10 ten

🖊️ **Write it!** Now, write 10 as a number and a word.

Tenth School Bus

First, there were 5 people on the bus.
Then, 5 more get on.

How many people are on the bus now?

🔍 **Solve it!**

Now, there are ⬚ people on the bus!

Apply it!

Have ready:

- 1 empty ten-frame
- objects or counters (10)
- number word cards (zero–ten)
- something to write with

This section helps your child use their number skills in simple problem-solving.

Each page includes:

- __Write it:__   Writing number words

- Solve it: Finding missing numbers

These activities connect hands-on learning (like using objects and pictures) with written maths and help build number confidence.

Talking together about what you see and do is a big part of the learning too. For example, if your child puts 6 objects on a ten-frame and then adds 4 more, you might ask:

How many
do you have
now?

I see
6 and 4 more!
How could we
count them
together?

These chats help your child make sense of the numbers, and grow their confidence too!

Iris has nine apples and picks one more.
How many apples does Iris have in total?

 **Build it!**

First, place 9 objects on a ten-frame. Add 1 more.

Draw it!

Then, draw 9 dots on the ten-frame, and add 1 more.

Write it!

Now, Iris has _ _ _ _ _ _ _ _ _ _ apples altogether.

Solve it!

9 and 1, that makes

James has eight carrot sticks on his plate.
He eats two.
How many carrot sticks does James have now?

 **Build it!**

First, place 8 objects on a ten-frame. Remove 2.

Draw it!

Then, draw 8 dots on the ten-frame. Cross out 2.

Write it!

Now, James has _________ carrot sticks.

 **Solve it!**

8 take away 2, that makes ⬚

Phoebe adds eight blueberries and two strawberries to her yogurt.
How many toppings does Phoebe have?

🔨 Build it!

First, add 8 and 2 on a ten-frame and find the total.

✏️ Draw it!

Then, draw 8 dots on the ten-frame. Add 2 more.

✏️ Write it!

Now, Phoebe has _ _ _ _ _ _ _ _ _ toppings altogether.

🔍 Solve it!

8 and 2, that makes ☐

Bobby opens a pod with eight peas inside.
He eats three.
How many peas are left?

🔨 Build it!

First, place 8 objects on a ten-frame. Remove 3.

✏️ Draw it!

Then, draw 8 dots on the ten-frame. Cross out 3.

🖊️ Write it!

Now, Bobby has _ _ _ _ _ _ _ _ _ peas.

🔍 Solve it!

8 take away 3, that makes [ ]

Ravi puts seven blueberries and three strawberries in his smoothie.
How many pieces of fruit are in Ravi's smoothie?

🔨 Build it!

First, place 7 objects on a ten-frame, then add 3 more.

✏️ Draw it!

Then, draw 7 dots on the ten-frame. Add 3 more.

🖊️ Write it!

Now, Ravi has _ _ _ _ _ _ _ _ _ pieces of fruit in total.

🔍 Solve it!

7 and 3, that makes

Annabelle has eight strawberries.
She gives four to a friend.
How many does Annabelle have now?

 **Build it!**

First, place 8 objects on a ten-frame. Remove 4.

Draw it!

Then, draw 8 dots on the ten-frame. Cross out 4.

Write it!

Now, Annabelle has _ _ _ _ _ _ _ _ _ strawberries.

 **Solve it!**

8 take away 4, that makes

Maya buys six oranges and four bananas. How many fruits does Maya have altogether?

 Build it!

First, place 6 objects on a ten-frame, then add 4 more.

Draw it!

Then, draw 6 dots on the ten-frame. Add 4 more.

Write it!

Now, Maya has _________ fruits in total.

 Solve it!

6 and 4, that makes

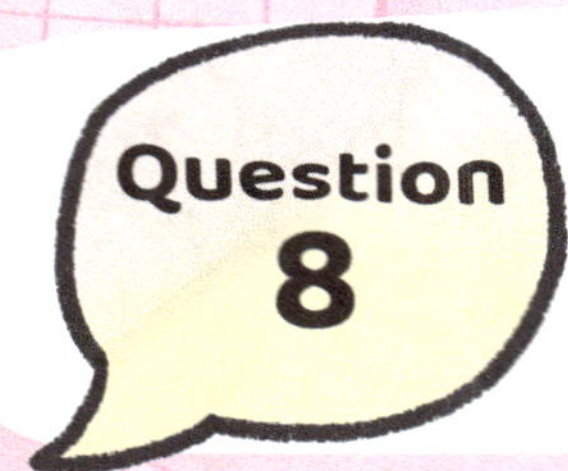

Luca has seven cherry tomatoes on his plate.
He eats three.
How many tomatoes are left?

🔨 Build it!

First, place 7 objects on a ten-frame. Remove 3.

✏️ Draw it!

Then, draw 7 dots on the ten-frame. Cross out 3.

🖊️ Write it!

Now, Luca has _________ tomatoes.

🔍 Solve it!

7 take away 3, that makes ☐

Louie picks five carrots and five potatoes from his garden.

How many vegetables does Louie have?

🔨 Build it!

First, place 5 objects on a ten-frame, then add 5 more.

✏️ Draw it!

Then, draw 5 dots on the ten-frame. Add 5 more.

🖊️ Write it!

Now, Louie has _ _ _ _ _ _ _ _ _ vegetables altogether.

⬤ Solve it!

5 and 5, that makes

There are six melon slices on Sofia's plate.
She eats two.
How many melon slices are left?

🔨 Build it!

First, place 6 objects on a ten-frame, then remove 2.

✏️ Draw it!

Then, draw 6 dots on the ten-frame. Cross out 2.

🖊️ Write it!

Now, Sofia has _ _ _ _ _ _ _ _ _ melon slices.

🔍 Solve it!

6 take away 2, that makes

A bus starts with four passengers.
At the next stop, six more people get on.
How many passengers are on the bus now?

🔨 Build it!

First, place 4 objects on a ten-frame, then add 6 more.

✏️ Draw it!

Then, draw 4 dots on the ten-frame. Add 6 more.

🖊️ Write it!

Now, there are ____________ passengers on the bus.

🔍 Solve it!

4 and 6, that makes ☐

A plane had five passengers.
One gets off when it lands.
How many passengers are still on the plane?

 **Build it!**

First, place 5 objects on a ten-frame. Remove 1.

Draw it!

Then, draw 5 dots on the ten-frame. Cross out 1.

Write it!

Now, there are _ _ _ _ _ _ _ _ _ _ passengers on the plane.

Solve it!

5 take away 1, that makes

There are three cars in a car park.
Seven more cars arrive.
How many cars are in the car park altogether?

 **Build it!**

First, place 3 on a ten-frame, add 7 more, and find the answer.

Draw it!

Then, draw 3 dots on the ten-frame. Add 7 more.

Write it!

Now, there are _ _ _ _ _ _ _ _ _ cars in total.

Solve it!

3 and 7, that makes

A bike shop had ten bikes.
Then four were sold.
How many bikes are left?

🔨 Build it!

First, place 10 objects on a ten-frame, remove 4, and count the remaining bikes.

✏️ Draw it!

Then, draw 10 dots on the ten-frame. Cross out 4.

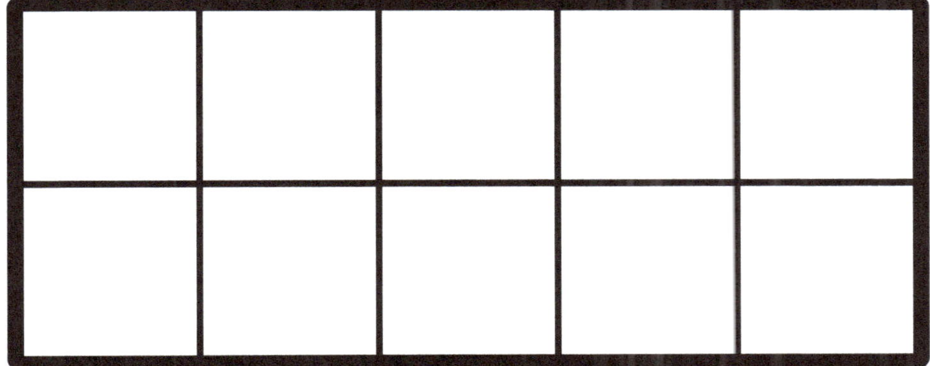

🖊️ Write it!

Now, there are _ _ _ _ _ _ _ _ _ bikes in the shop.

🔍 Solve it!

10 take away 4, that makes

A train has two carriages.
The driver adds eight more.
How many carriages does the train have now?

 **Build it!**

First, place with 2 objects on a ten-frame, add 8 more, and find the answer.

Draw it!

Then, draw 2 dots on the ten-frame. Add 8 more.

 **Write it!**

Now, the train has _ _ _ _ _ _ _ _ _ carriages.

 **Solve it!**

2 and 8, that makes [ ]

A rocket has nine astronauts.
Four astronauts go back to Earth.
How many astronauts are left inside the rocket?

 **Build it!**

First, place 9 objects on a ten-frame. Remove 4.

Draw it!

Then, draw 9 dots on the ten-frame. Cross out 4.

 **Write it!**

Now, there are _ _ _ _ _ _ _ _ _ astronauts in the rocket.

Solve it!

9 take away 4, that makes

There is one person in a hot air balloon.
Nine more climb in.
How many are in the balloon now?

🔨 Build it!

First, place 1 object on a ten-frame, add 9, and count the total.

✏️ Draw it!

Then, draw 1 dot on the ten-frame. Add 9 more.

✏️ Write it!

Now, there are _ _ _ _ _ _ _ _ _ people in the balloon.

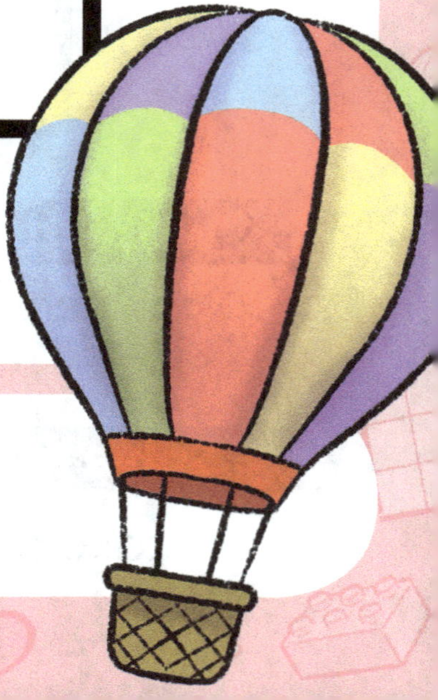

🔍 Solve it!

1 and 9, that makes

There are eight scooters at the park.
Four people ride away.
How many scooters are left?

🔨 Build it!

First, place 8 objects on a ten-frame, then subtract 4.

✏️ Draw it!

Then, draw 8 dots on the ten-frame. Cross out 4.

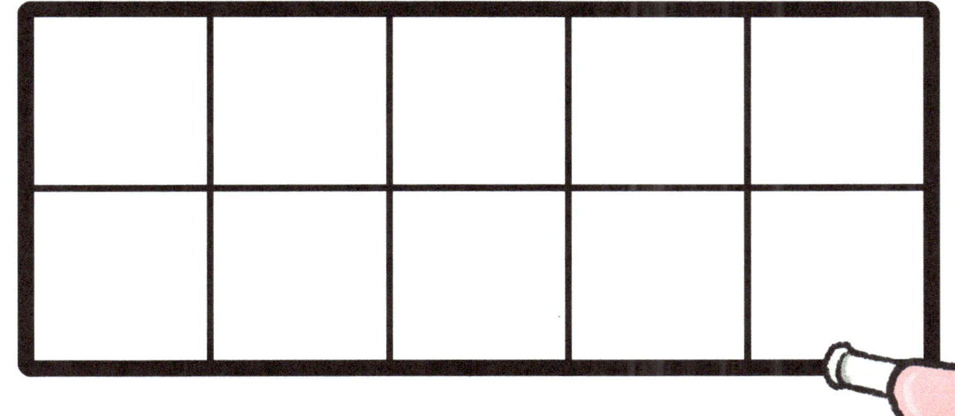

🖊️ Write it!

Now, there are ___________ scooters in the park.

🔍 Solve it!

8 take away 4, that makes [ ]

A tractor had an empty trailer.
At the farm, ten pigs walk on.
How many pigs are in the trailer altogether?

🔨 Build it!

First, place 0 objects on a ten-frame. Add 10 more.

✏️ Draw it!

Then, draw 0 dots on the ten-frame. Add 10.

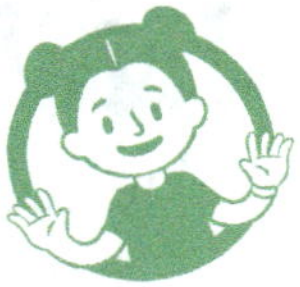

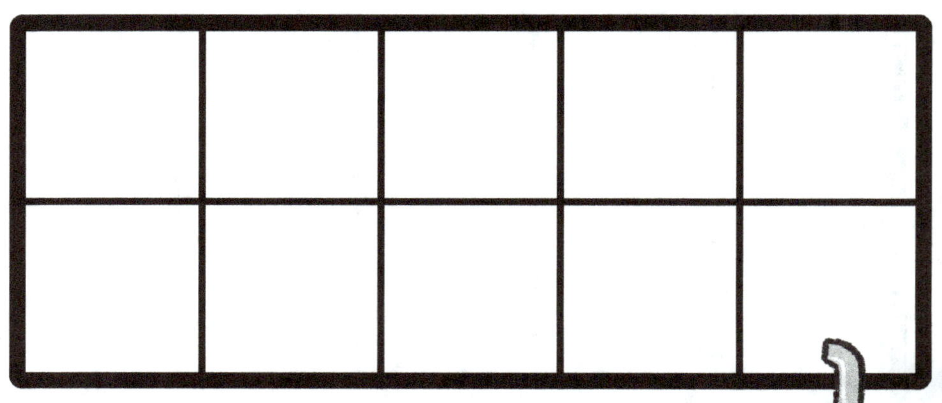

🖊️ Write it!

Now, there are _ _ _ _ _ _ _ _ _ pigs in the trailer.

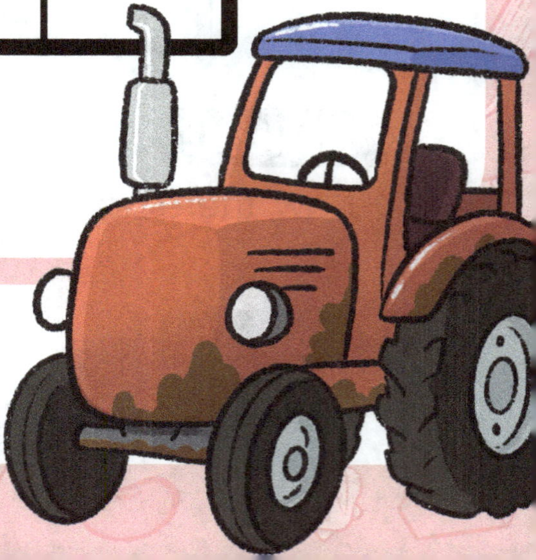

🔍 Solve it!

0 and 10, that makes [ ]

A post van started with seven parcels.
Four parcels were delivered.
How many parcels are on the van now?

🔨 Build it!

First, place 7 objects on a ten-frame, remove 4, and count what's left.

✏️ Draw it!

Then, draw 7 dots on the ten-frame. Cross out 4.

🖊️ Write it!

Now, there are _ _ _ _ _ _ _ _ parcels on the van.

🔍 Solve it!

7 take away 4, that makes 

See it! Say it!

Have ready:

- ten-frame flashcards (0-10)

- number cards (0-10)

- number word cards (zero-ten)

This section builds subitising skills:
the ability to see how many without counting.

Ten-frames help with this by showing amounts clearly.

Example: Seeing 3 counters and knowing it's 3.

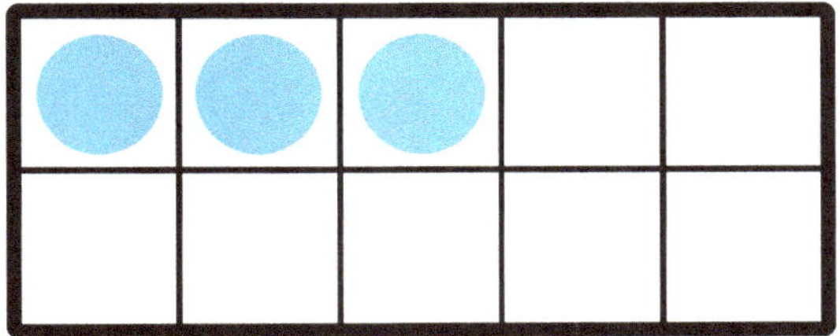

This helps children:

The Big Challenge!

 Shuffle the flashcards, flip one over, and ask:

Example: 4

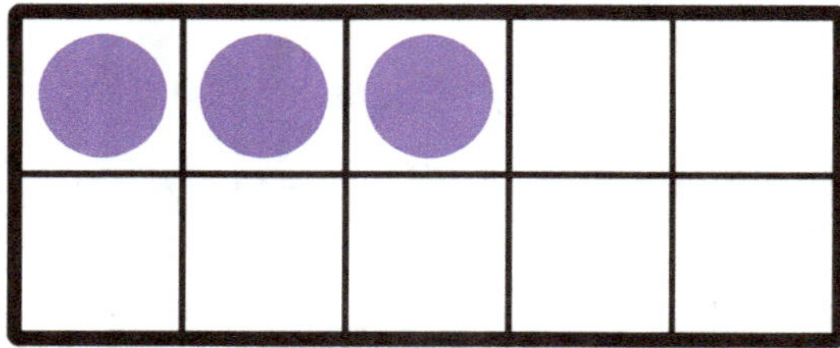

 Keep going until all cards are turned.

This section helps develop subitising skills by encouraging children to quickly identify how many objects are present without counting one by one.

The Bigger Challenge!

1 Have ready:

- Flashcards
- Number cards set out in order from 0-10
- Number words (zero-ten)

2 Ask your child to:

Say the number on the flashcard

Find the matching number card

Find the word
that matches

This activity helps with number recognition/order and reading skills.

The Biggest Challenge!

1 Pick a small number range (for example, 1-3).

2 Spread out flashcards, number cards, and word cards.

2 Ask your child to match the sets of three.

Add a timer for extra fun!

This activity builds:

- Visual recognition
- Number word understanding
- Counting and matching skills

Answers Explored

 **Number Stories and Patterns**

In the 'Draw it!' section, number stories are shown in sets to help your child spot patterns and connections. Each set follows a similar structure with small changes.

By looking at what's the same and what's different, children build a stronger understanding of how numbers work – and how they can be used in different ways.

Number Sentence Set	What's the Same?	What's Different?	Why It Matters
3 + 4 = 7 4 + 3 = 7	Both add up to 7	The numbers are in a different order	Shows that numbers can be added in any order (this is called "commutativity")
6 − 3 = 3 6 − 2 = 4	Both start with 6	A different number is taken away	Helps children compare and see how changing one part changes the result
6 + 4 = 10 10 − 4 = 6	Uses the same numbers: 6, 4, and 10	One adds, one subtracts	Helps children see the link between addition and subtraction
9 − 3 = 6 8 − 2 = 6	Both give the answer 6	Start and take-away numbers are different	Shows that there's more than one way to get the same answer
2 + 3 = 5 1 + 4 = 5	Both make 5	Different number sets	Helps children learn number bonds to 5 (sets of numbers that go together to make 5)

 Spotting Patterns through Number Stories

The 'Write it!' activities use bus-themed number stories to help your child spot patterns and build number sense.

Once your child has finished a set of stories, you can write out each set of number sentences and talk about them together.

These little changes help children notice what stays the same and what changes in maths.

Seeing these patterns helps your child understand how numbers grow, shrink, and connect to each other.

Number Sentence Set	What's the Same?	What's Different?	Why It Matters
Doubling Numbers 1 + 1 = 2 2 + 2 = 4 3 + 3 = 6 4 + 4 = 8 5 + 5 = 10	Each number is added to itself. These are called doubles.	The number being doubled changes (1, then 2, then 3…).	Doubles follow a clear pattern—each total is 2 more than the one before. Learning doubles helps your child remember number facts quickly and makes other adding easier later on (like 3 + 4 is just one more than 3 + 3).

Number Sentence Set	What's the Same?	What's Different?	Why It Matters
Halving Through Subtraction 2 – 1 = 1 6 – 3 = 3 10 – 5 = 5	Each time, we take away half of the number.	The numbers get bigger, but the pattern stays the same.	This helps your child understand what halving means. It's a step toward sharing, dividing, and thinking flexibly about numbers.
Taking Away One 8 – 1 = 7 10 – 1 = 9	Each time, we are taking away just one.	The starting number changes.	This helps children understand what happens when we take away one—we go back one number. It builds confidence with counting backwards and knowing "one less," which is important for subtraction and quick mental maths.

In this section, your child explores how numbers work through simple addition and subtraction stories.

These activities help your child spot patterns and make connections: important steps for building number sense and confidence.

After completing each set of problems, you can write the number sentences out and talk together about what's the same, what's different, and what they notice. These little chats help children think flexibly and build mental maths skills.

Number Sentence Set	What's the Same?	What's Different?	Why It Matters
Number Bonds to 10 (Adding) 9 + 1 = 10 8 + 2 = 10 7 + 3 = 10 6 + 4 = 10 5 + 5 = 10 4 + 6 = 10 3 + 7 = 10 2 + 8 = 10 1 + 9 = 10 0 + 10 = 10	All of the sets add up to 10. These are called number bonds to 10.	The numbers that make up the total change. Sometimes it's 9 + 1, sometimes it's 5 + 5 but they all equal 10.	Children start to see a pattern. As one number gets bigger, the other gets smaller. Knowing number bonds well helps your child add and subtract more confidently and solve problems faster later on.

Number Sentence Set	What's the Same?	What's Different?	Why It Matters
Group 1: Same starting number 8 – 2 = 6 8 – 3 = 5 8 – 4 = 4	Each one starts with 8.	The number we take away gets bigger, and the answer gets smaller.	This shows your child that subtracting more means having less. It's a clear pattern that helps them understand what subtraction means.
Group 2: Same answer (difference) 7 – 3 = 4 6 – 2 = 4 5 – 1 = 4	Each one equals 4.	Both the starting number and the number taken away change.	This shows that there's more than one way to make the same answer, which builds flexible thinking.
Group 3: Same number taken away 10 – 4 = 6 9 – 4 = 5 8 – 4 = 4 7 – 4 = 3	Each one takes away 4.	The starting number goes down, and so does the answer.	This helps children understand subtraction as taking away, and how the starting number affects the answer. It supports quick thinking with numbers.

Counter cut-outs

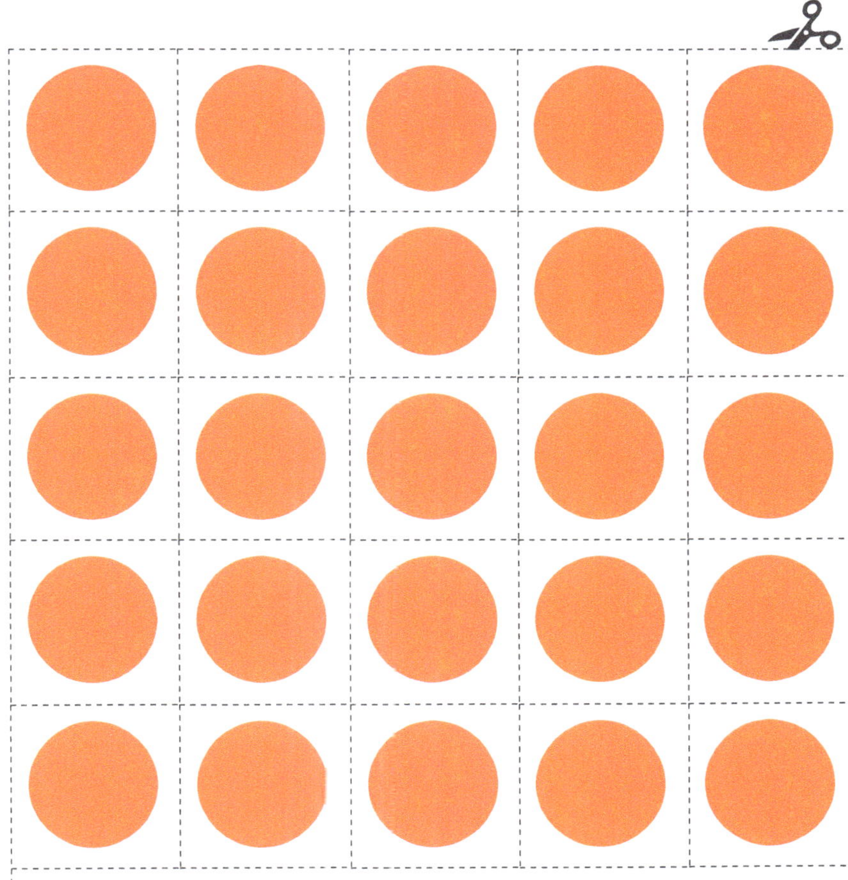

Ten-frame templates

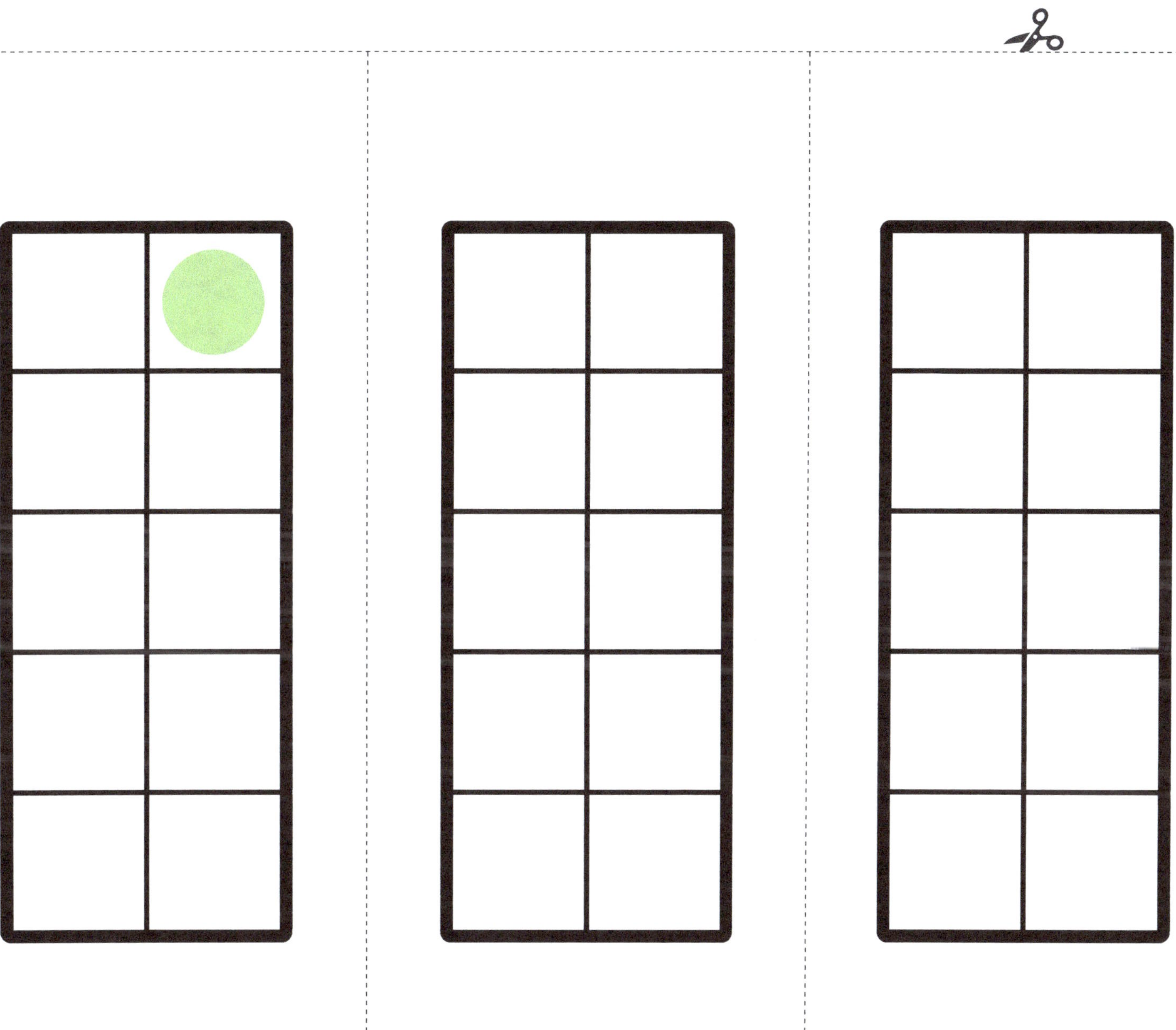

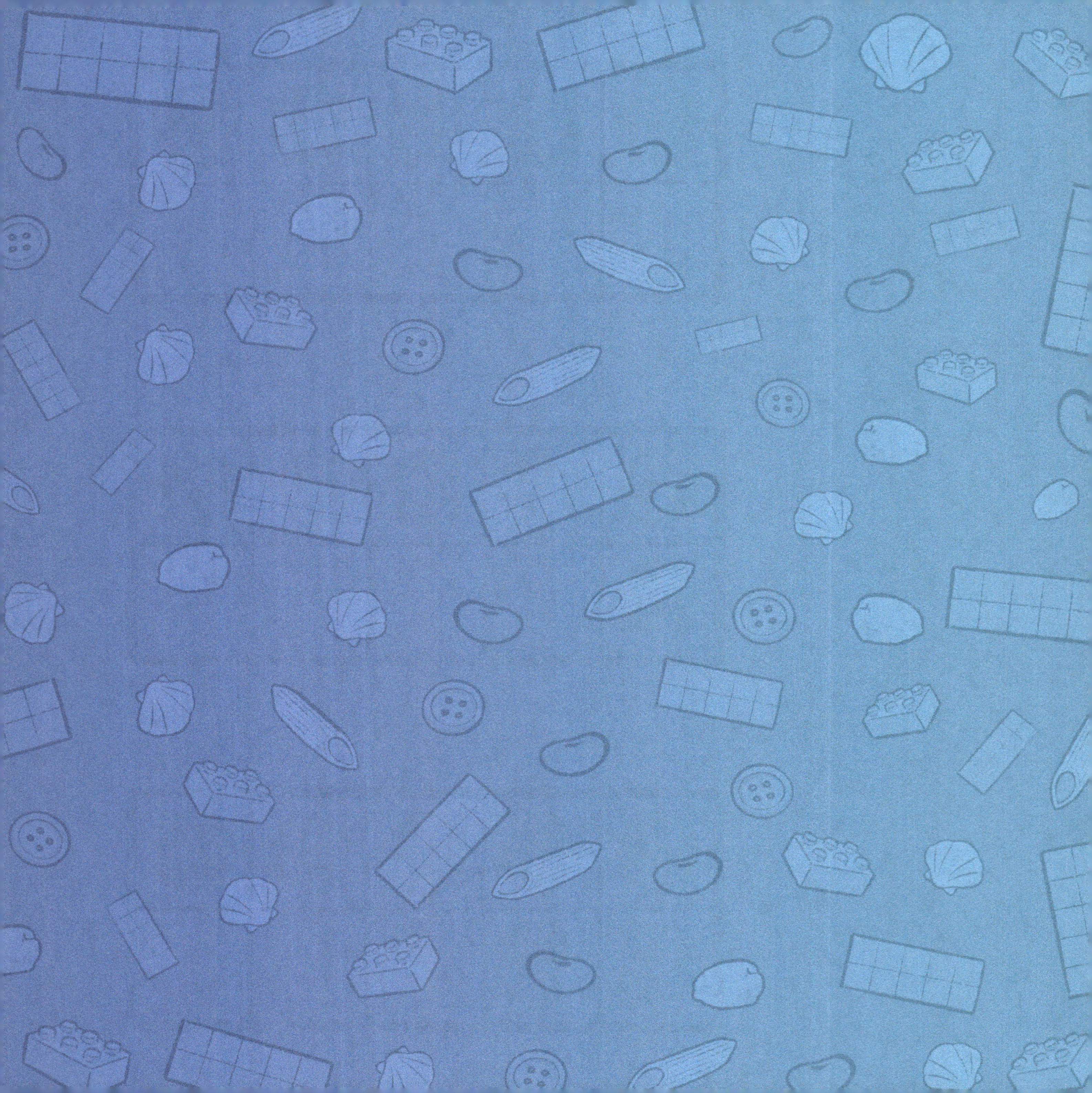

zero

one

two

three

four

five

six

seven

eight

nine

ten

0

1

2

3

4

5

6

7

8

9

10

Discover, Watch, Play

Workshops by Number+

https://www.numberplus.co.uk/

NRICH

https://nrich.maths.org/k/

Interactive Number Frames by Math Learning Centre

https://apps.mathlearningcenter.org/number-frames/

Scan the QR code
to leave a review

If you found this book valuable, please consider sharing your experience through **an honest rating or review.**

When readers share their thoughts on Amazon, it helps the platform understand who benefits most from this content, connecting the book with more educators just like you.

You can **leave a review** in under a minute by scanning this code with your phone camera.

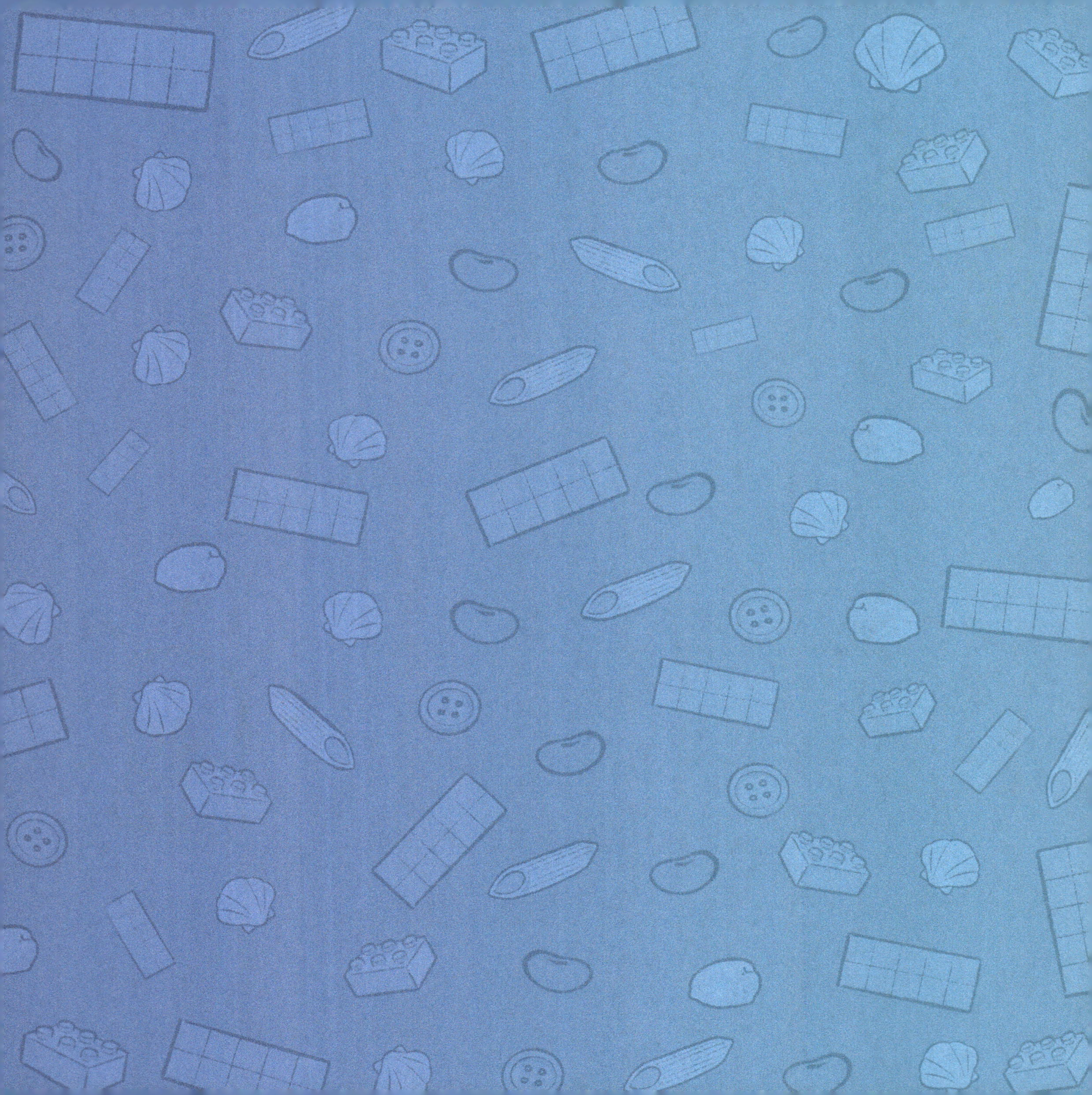